A LOOK AT CHEMISTRY

THE pH SCALE

BY MARY GRIFFIN

Gareth Stevens PUBLISHING

CRASHCOURSE

Please visit our website, www.garethstevens.com. For a free color catalog of all our high-quality books, call toll free 1-800-542-2595 or fax 1-877-542-2596.

Library of Congress Cataloging-in-Publication Data

Names: Griffin, Mary, 1978- author.
Title: The pH scale / Mary Griffin.
Description: New York : Gareth Stevens Publishing, [2019] | Series: A look at chemistry | Includes index.
Identifiers: LCCN 2018020798| ISBN 9781538230152 (library bound) | ISBN 9781538231449 (pbk.) | ISBN 9781538233320 (6 pack)
Subjects: LCSH: Hydrogen-ion concentration--Juvenile literature. | Acids--Juvenile literature. | Bases (Chemistry)--Juvenile literature.
Classification: LCC QD562.H93 G75 2019 | DDC 541/.3728--dc23
LC record available at https://lccn.loc.gov/2018020798

First Edition

Published in 2019 by
Gareth Stevens Publishing
111 East 14th Street, Suite 349
New York, NY 10003

Designer: Reann Nye
Editor: Therese Shea

Photo credits: Series art Marina Sun/Shutterstock.com; cover, p. 1 Rattiya Thongdumhyu/Shutterstock.com; p. 5 Julia700702/Shutterstock.com; p. 7 santol/Shutterstock.com; p. 9 Joerg Beuge/Shutterstock.com; p. 11 tab62/Shutterstock.com; p. 23 Stanislav Salamanov/Shutterstock.com; p. 25 Ming-Hsiang Chuang/Shutterstock.com; p. 27 David Gould/Photographer's Choice RF/Getty Images; p. 29 Cherries/Shutterstock.com; p. 30 (bleach) Beth Van Trees/Shutterstock.com; p. 30 (stomach acid) decade3d - anatomy online/Shutterstock.com; p. 30 (orange juice) stockcreations/Shutterstock.com; p. 30 (water) Bokeh Blur Background/Shutterstock.com; p. 30 (baking soda) EKramar/Shutterstock.com; p. 30 (black coffee) piikcoro/Shutterstock.com; p. 30 (ammonia solution) alenka2194/Shutterstock.com.

Printed in the United States of America

CPSIA compliance information: Batch #CW19GS: For further information contact Gareth Stevens, New York, New York at 1-800-542-2595.

CONTENTS

Words in the glossary appear in **bold** type the first time they are used in the text.

IT'S AN ACID!

Have you ever sucked on a lemon? You probably made a funny face because it was sour! Lemon juice tastes sour because it contains acids. Acids are **chemicals**. Some acids, such as those in lemon juice, are found in nature. Others are made by people.

MAKE THE GRADE

Your stomach produces a strong acid that breaks down the food you eat! It's called hydrochloric acid.

Acids can be weak or strong. Weak acids are in some foods you eat. For example, lactic acid is found in yogurt. Strong acids can be too strong to even touch. Sulfuric acid can break down metals!

MAKE THE GRADE

Sulfuric acid is used in some kinds of car **batteries**.

BITTER BASES

Bases are another kind of chemical. Bases often taste bitter and feel soapy when they're wet. They can be weak or strong, like acids can be. Baking soda, or sodium bicarbonate, is a weak base used in some foods.

MAKE THE GRADE

When baking soda is combined with an acid, gases are produced. This makes **dough** rise!

Strong bases aren't safe to eat or even touch. Lye, also called sodium hydroxide, is a base that breaks down animal and plant parts. It's used to make soap and paper. A base that **dissolves** in water may be called an alkali.

MAKE THE GRADE

A **substance** that isn't acidic or basic is called neutral.

PRESENTING THE pH SCALE

Scientists created the pH scale to **classify** acids and bases. The scale ranges from 0 to 14. Liquid **solutions** with a pH less than 7 are called acidic. Liquids with a pH greater than 7 are called basic.

THE pH SCALE

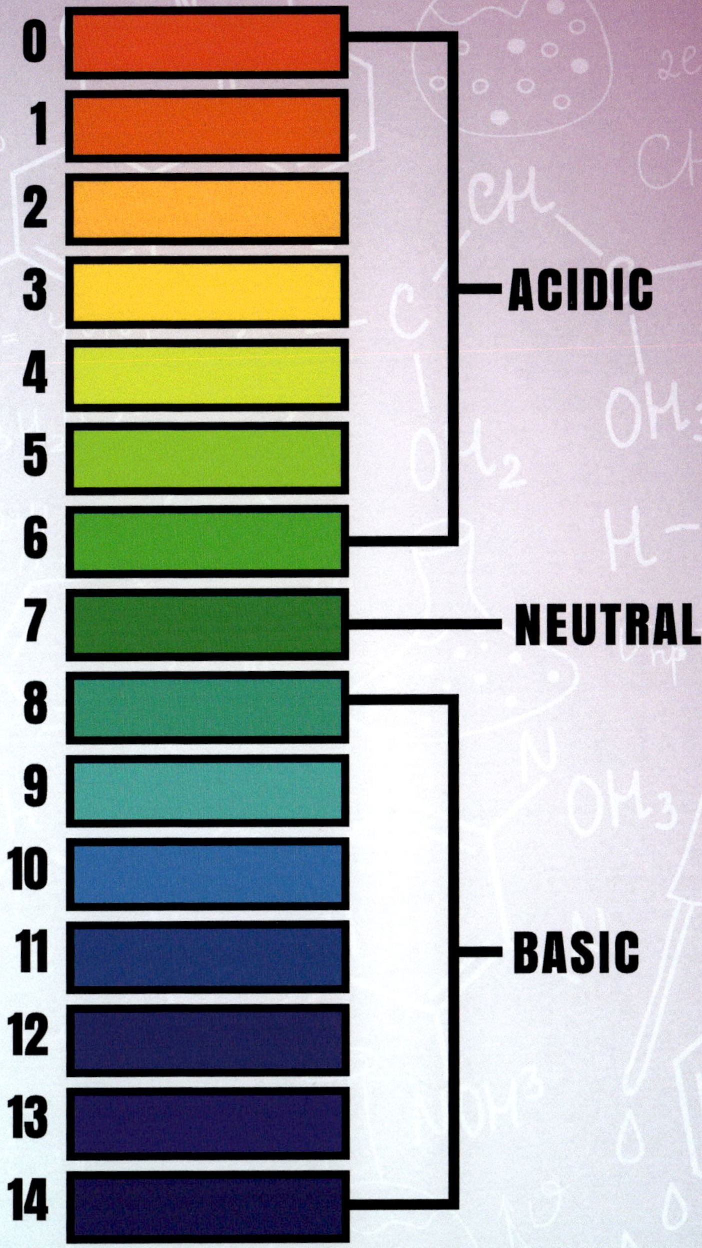

MAKE THE GRADE

The lower the pH number, the greater the acidity of the substance. The higher the number, the more basic.

INSIDE ACIDS AND BASES

Why is something an acid or a base? It has to do with the **particles** that make up matter, called atoms. Atoms are made of even tinier bits. An atom's center, or nucleus, contains protons and neutrons. Electrons travel around the nucleus.

OXYGEN ATOM

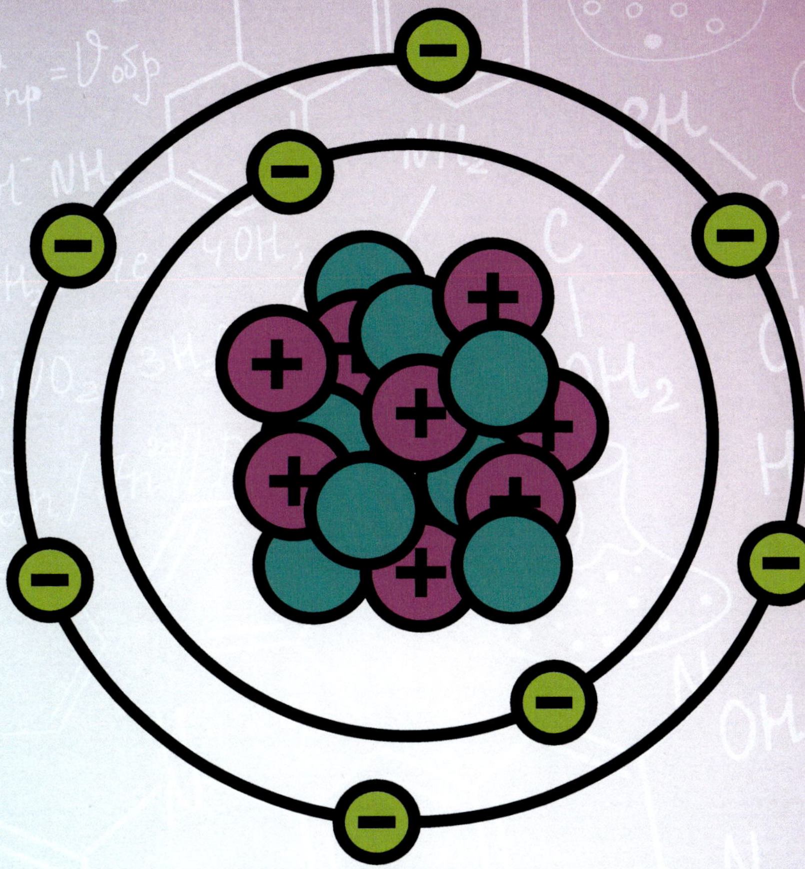

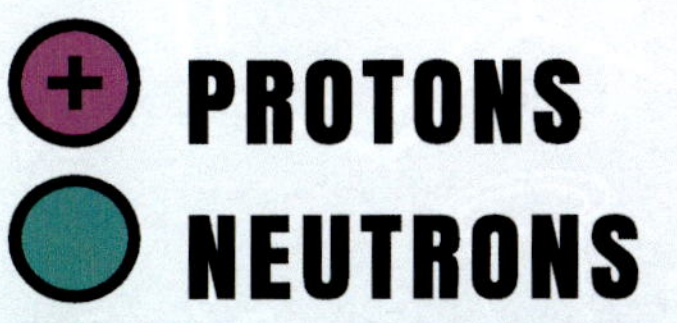

MAKE THE GRADE

Atoms make up the elements, substances that can't be broken down into other substances. Hydrogen and oxygen are examples of elements.

Protons and electrons have electric charges. Protons have a positive charge, and electrons have a negative charge. An atom usually has an equal number of protons and electrons so the charges balance. This means there's no overall electric charge.

PROTON
POSITIVE CHARGE

ELECTRON
NEGATIVE CHARGE

NEUTRON
NO CHARGE

MAKE THE GRADE

Neutrons have no electric charge.

Atoms can lose or gain electrons when they **react** with chemicals or crash into other atoms. A positive ion is an atom that has lost an electron. A negative ion is an atom that has gained an electron.

NEGATIVE ION
CHLORIDE

- 17 PROTONS
- 18 NEUTRONS
- 18 ELECTRONS

POSITIVE ION
SODIUM

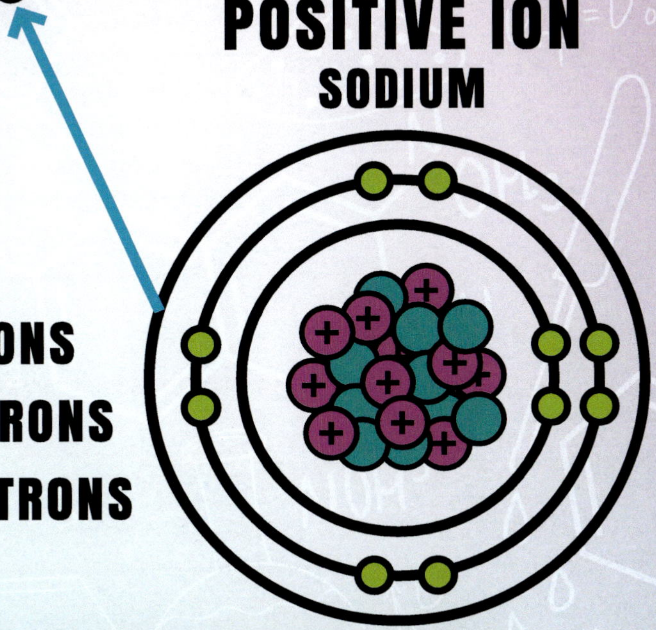

- 11 PROTONS
- 10 NEUTRONS
- 10 ELECTRONS

MAKE THE GRADE

A molecule is a group of atoms that are chemically bonded. Molecules that have lost or gained electrons are also called ions.

Pure water has an equal number of hydrogen ions and **hydroxide** ions. An acidic solution has more hydrogen ions than hydroxide ions. A basic solution has more hydroxide ions than hydrogen ions. A pH level is a measure of hydrogen ions in a solution.

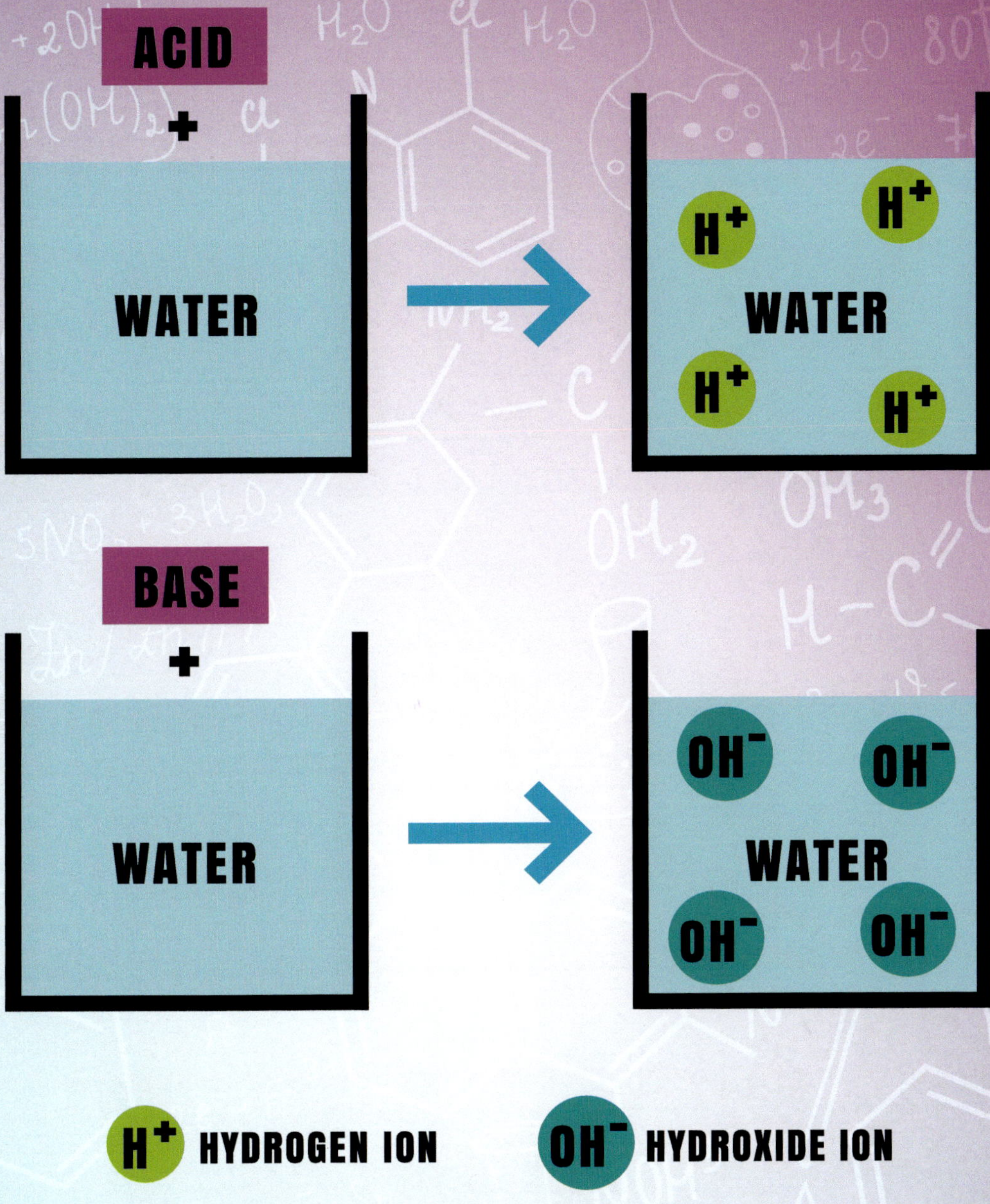

MAKE THE GRADE

When an acid dissolves in water, it produces hydrogen ions. When a base dissolves in water, it produces hydroxide ions.

A COLORFUL TEST

To test whether a substance is an acid or a base, scientists use indicators. Indicators indicate, or show, the presence of an acid or base, usually by changing color. The most well-known indicator is litmus paper.

MAKE THE GRADE

Litmus paper is treated with **dyes** from plantlike **organisms** called lichens.

Litmus paper can be different colors. Blue litmus paper turns red when a liquid solution is acidic. Red litmus paper turns blue when a solution is basic. Purple litmus paper turns red to indicate an acid and blue to indicate a base.

MAKE THE GRADE

The "p" in pH stands for *potenz*, which is the German word for "power." "H" is short for "hydrogen."

Litmus paper can't show how strong an acid or base is. An indicator called universal, or alkacid, indicator paper can turn many colors when dipped in a liquid. The color the paper turns is compared to the colors of the pH scale and indicates a substance's pH level.

MAKE THE GRADE

Water is neutral so it turns universal indicator paper green. This means water has a pH of 7.

NEUTRALIZE IT!

Mixing acids and bases neutralizes, or balances, their effects. For example, adding bases to acidic soil helps crops grow. Drugs called antacids contain bases that make you feel better when too much stomach acid makes you ill. Understanding acids and bases is useful!

MAKE THE GRADE

Acids and bases react when they come together. Together, they form salts!

EXPLORE THE pH SCALE

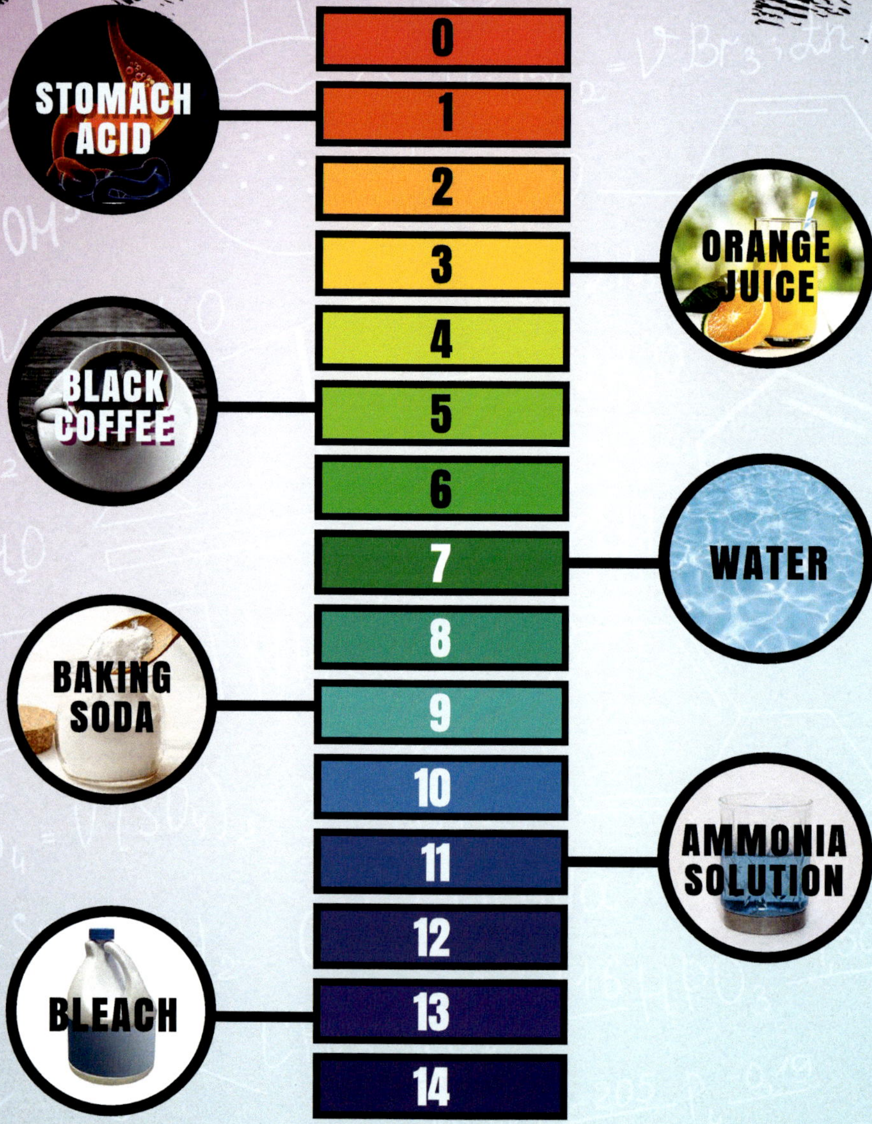

GLOSSARY

battery: a device that turns chemical energy into electricity

chemical: matter that can be mixed with other matter to cause changes

classify: to put something into a group based on shared features

dissolve: to mix with a liquid and become part of the liquid

dough: a mix of flour and water

dye: a substance used for changing the color of something

hydroxide: a negatively charged ion made up of oxygen and hydrogen atoms

organism: a living thing

particle: a very small piece of something

react: to change after coming in contact with another substance

solution: a liquid in which something has been dissolved

substance: a certain kind of matter

FOR MORE INFORMATION

BOOKS

Hartman, Eve, and Wendy Meshbesher. *Chemical Reactions.* Chicago, IL: Raintree, 2016.

Masoff, Joy, et al. *Oh, Ick! 117 Science Experiments Guaranteed to Gross You Out!* New York, NY: Workman Publishing, 2016.

WEBSITES

Acids and Bases
www.ducksters.com/science/acids_and_bases.php
Read more about acids, bases, and the pH scale.

Acids and Bases Are Everywhere
www.chem4kids.com/files/react_acidbase.html
Discover more about ions.

Publisher's note to educators and parents: Our editors have carefully reviewed these websites to ensure that they are suitable for students. Many websites change frequently, however, and we cannot guarantee that a site's future contents will continue to meet our high standards of quality and educational value. Be advised that students should be closely supervised whenever they access the internet.

INDEX